EXTINCT BEASTS

John Allan

Picture Credits
(abbreviations: t = top; b = bottom; m = middle; l = left; r = right; bg = background)

Shutterstock: 14-15bg, 28b; Catmando 16-17bg, 29bl; Daniel Eskridge 1bg, 6t, 8-9bg, 10-11bg, 28tl, 29tl, 30ml; Elenarts 20-21bg, ExpressionImage 4b; FOTOKITA 6b; Herschel Hoffmeyer 24-25bg, 28tr, 30br, 31bl; Kostiantyn Ivanyshen 22-23bg, 29mr; LukasKrbec 31tl; Lukas Uher 3b, 18-19bg; Orla 31mr; P Stock 2tl; rodos studio FERHAT CINAR 7t, 26-27bg; Warpaint 7m, 12-13bg.

Every effort has been made to trace the copyright holders and we apologise in advance for any unintentional omissions. We would be pleased to insert the appropriate credit in any subsequent edition of this publication.

Copyright © 2025 Hungry Tomato Ltd

First published in 2025 by Hungry Tomato Ltd
F15, Old Bakery Studios, Blewetts Wharf, Malpas Road, Truro, Cornwall,
TR1 1QH, UK.

No part of this publication may be reproduced, stored in a retrieval system, or transmitted in any form or by any means, electronic, mechanical, photocopying, recording, or otherwise, without prior written permission of the copyright owner.
A CIP catalogue record for this book is available from the British Library.

ISBN 9781835690796

Printed in China

Discover more at
www.hungrytomato.com

EXTINCT BEASTS

MEET THE WORLD'S MOST DANGEROUS PREHISTORIC CREATURES!

CONTENTS

Extinct Beasts	6	Allosaurus	20
Dilophosaurus	8	Spinosaurus	22
Velociraptor	10	Tyrannosaurus Rex	24
Giganotosaurus	12	Deinonychus	26
Troodon	14	The Fall of the Dinosaurs	28
Carcharodontosaurus	16	Fearsome Facts	30
Albertosaurus	18	Glossary & Index	32

Words in **BOLD** can be found in the glossary.

EXTINCT BEASTS

Which animal wins the title of deadliest dinosaur? This big question isn't as easy to answer as you might think! There's lots to consider...

THE DEADLIEST BEASTS

We have captured the most dangerous prehistoric creatures that roamed all corners of the Earth within the pages of this book...

HUNTERS BIG AND SMALL

The deadliest prehistoric creatures came in all shapes and sizes. Explore the top ten deadly animals going back in time, from huge Tyrannosaurus rex to small (but fast!) Velociraptors...

WHAT'S FOR DINNER?

Carnivores are animals that only eat meat, herbivores are animals that only eat plants and omnivores are animals that eat both! Most of the extreme **predators** we explore in this book are carnivores, which makes them the most dangerous of all!

DEADLY COUNTDOWN

All dinosaurs are ranked in order, from the 10th most deadly prehistoric predator, to the deadly dinosaur that takes the no.1 spot. It's not always the biggest that win!

WARNING
THINGS GET GRIM FROM HERE ON IN... TURN THE PAGES TO FIND OUT MORE!

DILOPHOSAURUS

Dilophosaurus lived during the early part of the Jurassic period. This carnivore had two bony ridges along the top of its skull. The first **fossil** remains of Dilophosaurus were discovered in 1942 in Arizona, USA.

CLEVER HUNTERS

Most scientists believe that Dilophosaurus hunted in packs, so it could attack animals much larger than itself. Dilophosaurus hunted by chasing down its **prey**. Powerful muscles in its back legs meant it could run quickly.

DEADLY PREDATOR

Although Dilophosaurus' jaws were packed with long, sharp teeth, they were quite weak! However, its claws were very sharp and very deadly! It had a slender build, and an unusually long tail for a predator.

FACT FILE

WEIGHT
410 kg
(900 lbs)

DIET
Carnivore

LOCATION
North America

LETHAL POWERS
Sharp claws, powerful back legs, and fast runner

DEADLY COUNTDOWN

NO.10

VELOCIRAPTOR

Velociraptor was a feathered dinosaur that lived towards the end of the Cretaceous period. It had a powerful combination of speed, aggression, and fearsome features – its small size is the only thing that stops it from being lower than 9 in the deadly countdown!

SPEEDY THIEF

The name Velociraptor means 'speedy thief'. Despite being one of the most ferocious dinosaurs found to date, they were only 2 metres (6.5 ft) long! One Velociraptor may not seem too deadly, but a pack of them definitely would have been!

FACT FILE

WEIGHT
Up to 45 kg
100 lbs

DIET
Carnivore

LOCATION
Asia

LETHAL POWERS
Long claws, 80 razor-sharp teeth, and fast runner

DEADLY COUNTDOWN

NO.9

SUPER SPEED

Velociraptors were fast-running dinosaurs, reaching speeds of up to 35 miles (65 km/h). They hunted in packs, often jumping onto the back of their prey using their sharp claws to hold on! A Velociraptor typically had 80 teeth, designed for easily ripping and tearing flesh.

GIGANOTOSAURUS

Giganotosaurus was one of the largest carnivores that ever walked the Earth. It is also one of the most mysterious of the meat-eating dinosaurs because it was only discovered in 1993.

THE GIANT

Giganotosaurus hunted by charging at its prey with its jaws wide open... scary! They attacked large herbivores using this technique, sometimes reaching speeds of 15 mph (24 km/h) – that's quick for such a big dinosaur!

DEADLY BUT DUMB

Giganotosaurus had jaws crammed with sharp, pointed teeth. They also had long tails that would have helped them stay balanced and turn quickly when running. They were fierce creatures but, despite being large in size, they had very small brains!

FACT FILE

WEIGHT
Up to 8,160 kg (18,000 lbs)

DIET
Carnivore

LOCATION
South America

LETHAL POWERS
Large sharp teeth, fast runner, and giant in size

DEADLY COUNTDOWN
NO.8

TROODON

Troodon was a small dinosaur that lived during the Cretaceous period. The first Troodon fossil was discovered in 1855 in North America. Despite being small, it may have been the brainiest animal of its time, placing the Troodon at number seven on the deadly countdown.

OSTRICH OR TROODON?
In terms of body shape, the Troodon was similar to an ostrich! It had long legs and a light body weight, meaning it could run very quickly. It is believed that the Troodon was covered in feathers, too.

EXCELLENT EYES

Troodons had 120 curved teeth, perfect for slicing through flesh. They also had large eyes that gave them excellent vision, helping them spot and chase small **mammals**, lizards, and snakes. It also meant they could hunt in the dark, allowing them to pounce on unsuspecting prey!

FACT FILE

WEIGHT
Up to 50 kg
(110 lbs)

DIET
Omnivore

LOCATION
Asia and
North America

LETHAL POWERS
Fast runner, super night vision, and curved sharp teeth

DEADLY COUNTDOWN

NO.7

CARCHARODONTOSAURUS

During the middle part of the Cretaceous period, this dinosaur was the top predator in North Africa. They were one of the largest dinosaurs, and often picked fights with dinosaurs bigger than themselves. The first Carcharodontosaurus fossil was found in 1927 in Algeria, Africa.

SNEAKY HUNTERS

It is believed that a Carcharodontosaurus was a sneaky hunter, hiding and waiting until it could launch a surprise attack on its prey. It had sharp teeth and wide, powerful jaws that were the size of a human!

FACT FILE

WEIGHT
Up to 6,350 kg (14,000 lbs)

DIET
Carnivore

LOCATION
North Africa

LETHAL POWERS
Powerful jaws, sharp teeth, and giant in size

SLOW BUT POWERFUL

Carcharodontosaurus were big and fierce to look at, but they were also very slow movers. They relied on their power and weight to hunt, rather than speed. If the prey spotted a Carcharodontosaurus in time, they would probably be able to outrun it!

DEADLY COUNTDOWN

NO.6

17

ALBERTOSAURUS

Albertosaurus was a slim, saw-toothed predator that hunted large plant-eating dinosaurs. It lived towards the end of the Cretaceous period. The first Albertosaurus fossil was found in 1884 in Canada.

MYSTERIOUS HUNTERS

Albertosaurus had a big head, with large, powerful jaws that were filled with around 60 razor-sharp teeth. Their eyes were positioned on the side of their head, which may have made it difficult for them to hunt – predators see better when their eyes are at the front.

FACT FILE

WEIGHT
Up to 2,495 kg
(5,500 lbs)

DIET
Carnivore

LOCATION
North America

LETHAL POWERS
Powerful jaws, fast runner, and sharp teeth

DEADLY COUNTDOWN

NO.5

DEADLY AND SCARY
This was still a very scary dinosaur! Its large size and fast speed made it a deadly hunter. It's believed the Albertosaurus could run as fast as 19 mph (30 km/h), which is very quick for a dinosaur of that size!

19

ALLOSAURUS

Allosaurus was one of the largest predators on Earth. It lived between the end of the Jurassic period and the beginning of the Cretaceous period. The first Allosaurus fossil was found in 1877 in North America.

TOP PREDATOR

Allosaurus was a predator with no natural enemies. It may have hunted in packs, using the large claws on its forelimbs to grab prey. This dinosaur would attack plant-eating dinosaurs as well as other predators, too!

FACT FILE

WEIGHT
Up to 2,720 kg
(6,000 lbs)

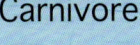

DIET
Carnivore

LOCATION
Africa, Europe, and
North America

LETHAL POWERS
Large claws, sharp teeth,
and a fierce fighter

DEADLY COUNTDOWN

NO.4

FIERCE OR FAKE?
They had sharp, long teeth. However, they were fragile and broke off easily. An Allosaurus could reach speeds of up to 12 mph (19 km/h), but it was unlikely to have enough **stamina** for a long chase.

21

SPINOSAURUS

Spinosaurus was the largest known carnivorous dinosaur that ever lived, and it had a deadly reputation! This fierce predator lived during the Cretaceous period. The first Spinosaurus fossil was discovered in Egypt in 1912.

LARGE CREST

The Spinosaurus had a slim body and long forelimbs, meaning it may have walked on all fours some of the time. It had a strange, tall crest on its back made of long spines, that is thought to have helped control its body temperature.

FACT FILE

WEIGHT
Up to 17,240 kg
(38,000 lbs)

DIET
Carnivore

LOCATION
Africa

LETHAL POWERS
Snapping jaws, heavy weight, and sharp teeth

DEADLY COUNTDOWN

NO.3

FISH SNATCHER
This dinosaur had long, narrow jaws, just like a crocodile! The shape of its jaws and teeth suggests that Spinosaurus fed mainly on fish from rivers and lakes.

TYRANNOSAURUS REX

This is the most famous of all the dinosaurs and was amongst the biggest land predators that ever lived! It is often known as T. rex because 'rex' means 'king' in Latin. The first fossil was discovered in 1905.

TERRIFIC TEETH

This fierce dinosaur was extremely scary! It would probably have charged at its prey with its mouth wide open, showing off its impressive set of teeth. It could tear off huge chunks of flesh with a single bite!

FACT FILE

WEIGHT
Up to 6,800 kg
(15,000 lbs)

DIET
Carnivore

LOCATION
Asia and
North America

LETHAL POWERS
Strong muscles, sharp teeth, and powerful jaws

DEADLY COUNTDOWN
NO.2

SUPER SENSES
Tyrannosaurus had massive jaws with very powerful muscles that could bite through even the biggest of bones! They were very effective predators, thanks to their great sense of smell and forward-facing eyes. Tyrannosaurus probably stalked lots of plant-eating dinosaurs and picked off the weakest members.

DEINONYCHUS

It's less well-known than T. rex, but Deinonychus was the ultimate dinosaur predator. It lived during the early part of the Cretaceous period, and takes the number one spot in the countdown, simply for its ferocious hunting tactics!

KNIFE-LIKE TEETH

Deinonychus had powerful jaw muscles and curved teeth. Each tooth could cut through skin and muscle like the blade of a knife! Because it was so small and light, Deinonychus was a very **agile** and fast-running predator. It could easily chase down dinosaurs much larger than itself!

PACK HUNTERS

Just one of these dinosaurs was scary enough, but no animal stood a chance against a whole pack! Deinonychus hunted in large groups, often jumping on the back of their prey and tearing their flesh with both claws and teeth.

FACT FILE

WEIGHT
Up to 100 kg
(220 lbs)

DIET
Carnivore

LOCATION
North America

LETHAL POWERS
High stamina, sharp curved teeth, and strong limbs

DEADLY COUNTDOWN

NO.1

THE FALL OF THE DINOSAURS

These prehistoric predators were some of the most fearsome creatures to ever walk the Earth. However, they faced many challenges, and not just the threat of being eaten!

VELOCIRAPTOR

While Velociraptors were super-fast and aggressive hunters, this wasn't enough to outrun much larger predators. Huge dinosaurs, such as the Tarbosaurus, ate these speedy thieves for breakfast!

Carcharodontosaurus

This dinosaur was the top predator in North Africa, where they roamed the rainforests and plains! However, due to climate change, large-scale habitat loss caused the Carcharodontosaurus to die out.

T.REX

This ferocious dinosaur is one of the most famous predators to have ever lived. However, their huge jaws and powerful muscles were no match for the giant **asteroid** that hit the Earth 66 million years ago.

COLLISION COURSE

Some dinosaurs did survive this giant asteroid! However, the collision caused a large cloud of dust and ash to block out the sun. This meant that many plants died and as a result, the food chain collapsed. The surviving dinosaurs struggled to survive from the lack of food they could find.

FEARSOME FACTS

There are so many deadly and fearsome facts about each prehistoric predator. Here are some more that show just how impressive these dinosaurs really were...

DILOPHOSAURUS
The reason for the crest on top of this dinosaur's skull remains a mystery. Some scientists believe it could have been a sign of strength, or to help them attract a **mate**.

VELOCIRAPTOR
An entire fossil of a Velociraptor attacking a herbivore dinosaur was discovered in Mongolia in 1971.

GIGANOTOSAURUS
Despite its huge size, the Giganotosaurus had a very small brain. It was around the size of a cucumber!

TROODON
In the USA, a Troodon nesting ground was discovered with many eggs inside it. Some scientists believe that female Troodons laid around 2 eggs per day!

30

ALBERTOSAURUS
The remains of 26 Albertosaurus were uncovered together in Canada. This suggests that these predators lived in herds.

CARCHARODONTOSAURUS
This deadly predator was named after the scientific name for many sharks, Carcharodon. This dinosaur's teeth looked very similar to that of a great white shark!

SPINOSAURUS
It is believed that Spinosaurus were able to swim. However, because of their size and shape, they were most likely very awkward and slow in the water!

ALLOSAURUS
Allosaurus were constantly shedding and growing new teeth! There are so many that have been found that it's possible to buy them.

T.REX
Scientists believe that Tyrannosaurus rex were among other dinosaurs that were covered in feathers when they hatched!

DEINONYCHUS
Research suggests that this extreme predator had the same bite force as an alligator!

GLOSSARY

Agile – able to move quickly and easily.

Asteroid – a space rock.

Crest – a ridge on the head of an animal.

Cretaceous period - a period in time between 145 and 66 million years ago.

Fossil – evidence of past life that has turned to stone over time.

Jurassic period – a period in time between 201 and 145 million years ago.

Mammals – warm-blooded animals with a covering of hair on the skin and the ability to produce milk to feed their young.

Mate – one of a pair of animals that live or have babies together.

Predators – animals that live by attacking and killing other animals.

Prey – an animal hunted or caught for food.

Stamina – lasting strength and energy.

INDEX

A
Albertosaurus 18-19, 31
Allosaurus 20-21, 31

C
Carcharodontosaurus 16-17, 28, 31
Carnivore 7, 8-9, 10-11, 12-13, 14-15, 16-17, 18-19, 20-21, 22-23, 24-25, 26-27, 28-29, 30-31

D
Deinonychus 26-27, 31
Dilophosaurus 8-9, 30

G
Giganotosaurus 12-13, 30

H
herbivore (plant-eater) 7, 30

O
omnivore 7, 14-15

S
Spinosaurus 22-23, 31

T
Troodon 14-15, 30
Tyrannosaurus rex 6, 24-25, 29, 31

V
Velociraptor 6, 10-11, 28, 30